# Mastering the

# Magic Pyramid

**The secrets of the Pyraminx unlocked**

**Tom Werneck**

# Evans Brothers Limited

Published by Evans Brothers Limited,
Montague House, Russell Square,
London WC1B 5BX

First published 1981
© Wilhelm Heyne Verlag 1981

Original edition first published by Wilhelm Heyne Verlag,
Munchen.

Photographs: Arxon, Rodgau; Beinengraber, Hamburg; Meffert,
Hong Kong.

Set by Technical Editing Services
and printed in Spain
Printer ind. graf. sa Barcelona
D.L.B. 27806-1981

ISBN 0 237 45591 9

# Contents

The Publishers would like to thank David Singmaster for all his invaluable help in the production of this book.

# Foreword

The idea for the Pyraminx™ is almost ten years old. When I made my first pyramid, I did not intend it as a game or a toy. I was merely fascinated by the perfectly symmetrical shape. Its simple geometric form radiates an intense power and has a soothing influence. If you stroke the palm of your hand with the rounded peak of a small pyramid, you will slowly relax.

I started working on the pyramid with my brother, a versatile and gifted engineer. We began by building a wooden model. Once we had solved all of the construction problems, we put the completed pyramid in a drawer and forgot about it.

When Rubik's Cube™ captured the world's imagination, I realized that my pyramid could also challenge the intellect. I contacted Tom Werneck, who had worked with games for many years, and he was very enthusiastic. He agreed to write this book so that an easily understood solution to the pyramid might be published and made widely available.

Playing with the pyramid – whether one is solving it for the first time or creating patterns on its surfaces –

can be a satisfying and aesthetically pleasing experience. I hope that the readers of this book will find it so.

Uwe Mèffert
Hong Kong
August, 1981

# Chapter 1: Terminology and Turning Moves

Before you begin solving the Pyraminx, you must become very familiar with how the various parts are named and how they may be manipulated.

Each side is called a *face*. The pyramid has four faces.

Hold the pyramid as shown in the illustration, with

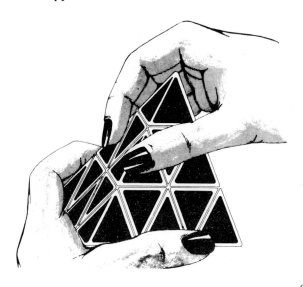

one full face and the face immediately to its left both visible. This is how you will normally hold the pyramid while you are performing the turning moves.

In order to show what happens as a result of each step in the solution sequence, many of the illustrations are drawn in such a way as to reveal all four faces of the pyramid at once. These are termed the *front face*, the *left face*, the *right face*, and the *bottom face*. The six edges of the pyramid are called the *left front edge*, the *right front edge*, the *back*

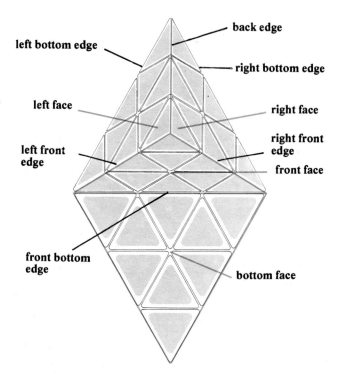

*edge,* the *front bottom edge,* the *left bottom edge,* and the *right bottom edge.*

Each face consists of three vertical rows. The top and middle rows together are called the *upper layer.*

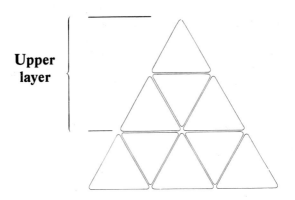

**Upper layer**

You must also learn to think in terms of layers that cross the pyramid at 60 degree angles. These are called the *right front layer,* the *left front layer,* and the *back layer.*

All four of the corner pieces are called *peaks.*

The illustrations which follow demonstrate all of the ways in which the layers and peaks of the pyramid may be turned.

Each diagram indicates that you are to turn a particular layer or peak 120 degrees (one third of the way around the pyramid). The mechanism will click into place when the turning move is completed.

# Turning the upper layer:

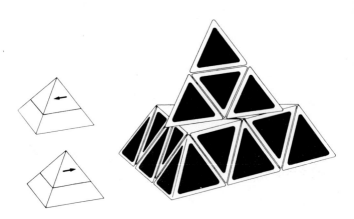

# Turning the right front layer:

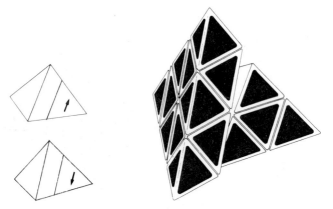

## Turning the left front layer:

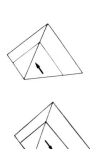

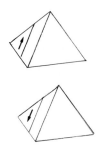

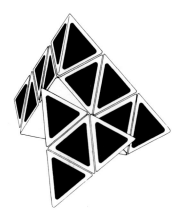

## Turning the back layer:

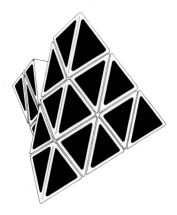

Turning the peaks:

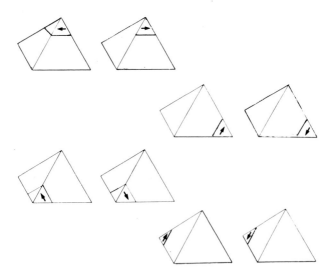

You will be solving the pyramid in three stages.

In Stage 1, you will correctly position the four peaks (colored grey in the illustration):

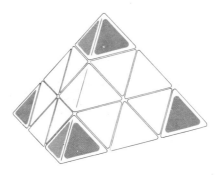

In Stage 2, you will correctly position the edge pieces (colored grey in the illustration):

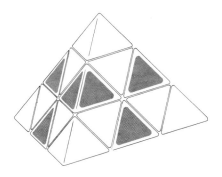

In Stage 3, you will fit the middle pieces into place (colored grey in the illustration):

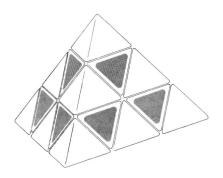

# Chapter 2:
# Solving the Pyraminx

*Note:* Before you begin working on the solution, take a few moments to play with your pyramid. When you are comfortable with how it feels in your hand and can turn it quickly and easily, you are ready to proceed.

IMPORTANT: Although it is common practice to lubricate the cube with petroleum jelly or silicone spray, it is *not* recommended that you do the same with the Pyraminx. The inner sphere on which the mechanism turns is made of a self-lubricating plastic. Further lubrication is not needed and may in fact have a harmful effect.

It is also common practice to take the cube apart and put it back together again. The Pyraminx, on the other hand, does not come apart easily. Attempting to separate the pieces may irreparably damage the mechanism.

### Stage 1: Positioning the Peaks

Begin by choosing one of the four colors on the pyramid – red, green, yellow, or blue. It doesn't matter which color you decide upon. For purposes of illustration, we have chosen to start with red and will present the solution sequence in those terms.

Locate the peak that does *not* contain the color red. Hold the pyramid by this peak. The other three peaks will each have a red surface.

Turn these three peaks until the red surfaces are all on the same face, as shown in the illustration below. This will be the red face.

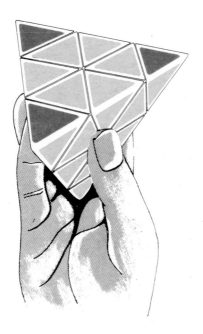

Now turn the pyramid over so that the red face is on the bottom. Turn the peak at the top until your pyramid looks like the one shown below. You have now determined the green, blue, and yellow faces of your pyramid and correctly positioned all four peaks.

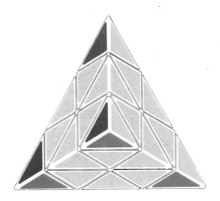

*Note:* Throughout the solution sequence, ignore all areas which are shaded light grey in the color illustrations. These will be taken care of in time.

## Stage 2: Positioning the Edge Pieces

**Step 1:** Positioning the green/red edge piece. During this step the green face is the front face and the red face is the bottom face.

Locate the green/red edge piece.

In order for this piece to be correctly positioned, it must be in the area shaded dark grey in the illustration. The green surface must be on the front face and the red surface on the bottom face.

**a** If the green/red edge piece is already correctly positioned, move on to Step 2.

**b** If the green/red edge piece is in the left front edge

with the red surface to the front, perform the following turning move:

The green/red edge piece is now in position.

You will notice that the left peak now shows a blue surface on the green face. Correct this by performing the following turning move:

This will reposition the peak.

IMPORTANT: You will need to reposition a peak following *each* turning move in the solution sequence.

**c** If the green/red edge piece is in the right front edge with the red surface to the front, perform the following turning move:

The green/red edge piece is now in position.

**d** If the green/red edge piece is in the left front edge with the green surface to the front, perform the following turning move:

Now go back to **c** and perform the move described there.

**e** If the green/red edge piece is in the right front edge with the green surface to the front, perform the following turning move:

Now go back to **b** and perform the move described there.

**f** If the green/red edge piece is in the back edge, move the upper layer so that the red surface comes to either the right front edge or the left front edge. Now go back to **b** or **c** and perform the move described there.

**g** If the green/red edge piece is in the left bottom edge, perform this turning move:

If necessary, move the upper layer so that the red surface of the green/red edge piece comes to the front face.

Depending on where the green/red edge piece now is – in the left front edge, or in the right front edge – go back to either **b** or **c** to complete the turning move sequence.

**h** If the green/red edge piece is in the right bottom edge, perform this turning move:

If necessary, move the upper layer so that the red surface of the green/red edge piece comes to the front face.

Depending on where the green/red edge piece now is – in the left front edge, or in the right front edge – go back to either **b** or **c** to complete the turning move sequence.

**i** If the green/red edge piece is in the front bottom edge but is incorrectly oriented – in other words, if the red surface is on the front face and the green surface is on the bottom face – perform either of the following turning move sequences:

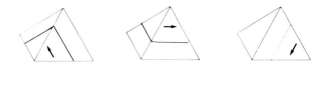

*or*

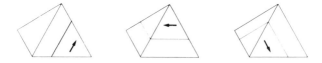

End this step by repositioning any peaks that may have been moved out of place.

**Step 2:** Positioning the blue/red edge piece. During this step, the blue face is the front face and the red face is the bottom face.

Locate the blue/red edge piece.

In order for this piece to be correctly positioned, it must be in the front bottom edge with the blue surface on the front face and the red surface on the bottom face.

**a** If the blue/red edge piece is already correctly positioned, move on to Step 3.

24

**b** If the blue/red edge piece is in the left front edge with the red surface to the front, perform the following turning move:

**c** If the blue/red edge piece is in the right front edge with the red surface to the front, you *cannot* simply turn the right front layer forward to position the piece because this will displace the green/red edge piece you positioned in Step 1.

The blue/red edge piece in the right front edge.

The green/red edge piece displaced as a result of incorrectly moving the blue/red edge piece.

Instead, move the upper layer to the left to bring the blue/red edge piece to the left front edge.

Now move the green/red edge piece *out of the way:*

Now move the blue/red edge piece back to the right front edge:

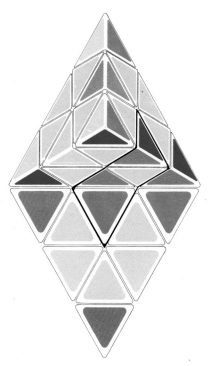

Now position the blue/red edge piece *and* reposition the green/red edge piece by means of the following turning move:

Here is the complete turning move sequence again:

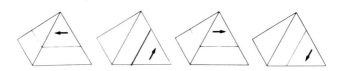

**d** If the blue/red edge piece is in the left front edge with the blue surface to the front, perform this turning move:

Now go back to **c** above and perform the sequence described there to correctly position the blue/red edge piece.

**e** If the blue/red edge piece is in the right front edge with the blue surface to the front, perform the following turning move.

Now go back to **b** to correctly position the blue/red edge piece.

**f** If the blue/red edge piece is in the back edge, move the upper layer so that the red surface comes to either the right front edge or the left front edge. Now go back to either **b** or **c** to correctly position the blue/red edge piece.

**g** If the blue/red edge piece is in the left bottom edge, perform this turning move:

If necessary, move the upper layer so that the red surface of the blue/red edge piece comes to the front face.

Depending on where the blue/red edge piece now is – in the left front edge or the right front edge – go back to either **b** or **c** to complete the turning move sequence and correctly position the piece.

**h** The blue/red edge piece *cannot* be in the right bottom edge, since this position is already occupied by the green/red edge piece.

**i** If the blue/red edge piece is on the front bottom edge but is incorrectly oriented – in other words, if the red surface is on the front face and the blue surface is on the bottom face – perform the following turning move sequence:

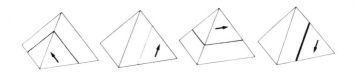

**Step 3:** Positioning the yellow/red edge piece. During this step, the yellow face is the front face and the red face is the bottom face.

In order for this piece to be correctly positioned, it must be in the front bottom edge with the yellow surface on the front face and the red surface on the bottom face.

**a** If the yellow/red edge piece is already correctly positioned, move on to Step 4.

If the yellow/red edge piece is not yet correctly positioned, you will find it in the left front edge, the right front edge, or the back edge. It is also possible for it to be positioned in the front bottom edge but incorrectly oriented – that is, with its colors turned around.

The right bottom edge and the left bottom edge are already occupied by correctly positioned edge pieces.

*Note:* At this stage of the solution an unusual pattern sometimes occurs. Before proceeding with Step 4, see *Special Case* on page 46.

**b** If the yellow/red edge piece is in the left front edge with the red surface to the front, you *cannot* simply move the left front face forward to position this piece, since that would displace the green/red edge piece.

Instead, move the upper layer to the right to bring the yellow/red edge piece to the right front edge:

Now move the green/red edge piece *out of the way:*

33

Now move the yellow/red edge piece back to the left front edge:

Finally, position the yellow/red edge piece *and* reposition the green/red edge piece by means of the following turning move:

Here is the complete turning move sequence.

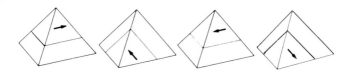

**c** If the yellow/red edge piece is in the right front edge with the red surface to the front, you *cannot* simply move the right front face forward to position this piece, since that would displace the blue/red edge piece.

Instead, perform the following sequence of turning moves:

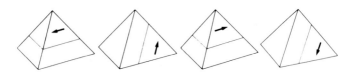

Note that this sequence is the mirror image of the one described in **b** above.

**d** If the yellow/red edge piece is in the left front edge with the yellow surface to the front, perform the following turning move:

Now go back to **c** to correctly position the yellow/red edge piece.

**e** If the yellow/red edge piece is on the right front edge with the yellow surface to the front, perform the following turning move:

Now go back to **b** to correctly position the yellow/red edge piece.

**f** If the yellow/red piece is in the back edge, turn the upper layer so that the red surface comes to either the right front edge or the left front edge. Now go back to either **b** or **c** to correctly position the yellow/red edge piece.

**g** If the yellow/red edge piece is in the front bottom edge but is incorrectly oriented – in other words, if the red surface is on the front face and the yellow surface is on the bottom face – perform the following turning move sequences:

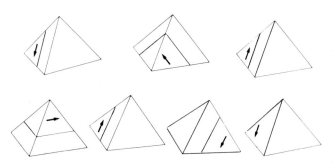

End this step by repositioning any peaks that may have been moved out of place.

Now turn your pyramid over. You will notice that all of the red edge pieces are correctly positioned, as shown in the illustration below:

If you turn your pyramid over again, you will see that the edge pieces along the red face are all correctly positioned.

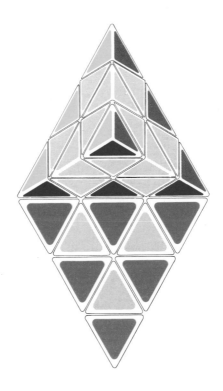

*Note:* At this stage of the solution an unusual pattern sometimes occurs. Before proceeding with Step 4, see *Special Case* on page 46.

**Step 4:** Positioning the remaining edge pieces. Turn the pyramid so that the red face is at the bottom and you are looking straight down at the top peak.

Now turn the upper layer until *one* of the remaining middle edge pieces is correctly positioned. In other words, both surfaces of the edge piece must match the colors of the adjacent peaks. Turn the pyramid so that the correctly positioned edge piece is in the back edge.

*Note:* If you are *very* lucky, turning the upper layer of the pyramid will correctly position all three of the remaining edge pieces at once, and you can go on to Stage 3.

In the illustration at the right, the blue/yellow edge piece has been correctly positioned. But the edge pieces indicated by the dark grey areas have not. To correct this, perform the following sequence of turning moves. The color illustrations beside the turning move diagrams show what will happen as a result of each turning move *if* you begin with the properly positioned edge piece – in this case, blue/yellow – in the back edge.

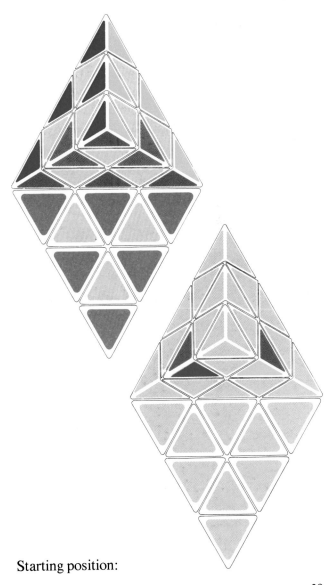

Starting position:

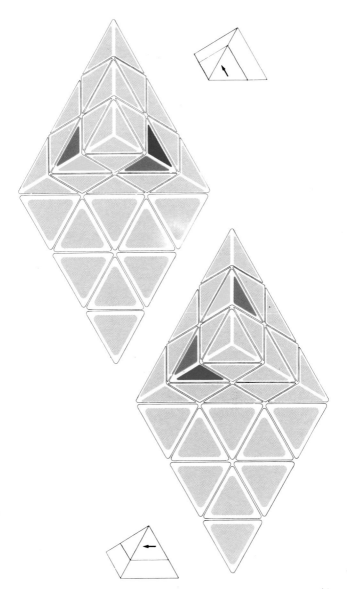

Here is the sequence again without the color illustrations. Use this sequence whenever you need to correct two of the edge pieces in the upper layer.

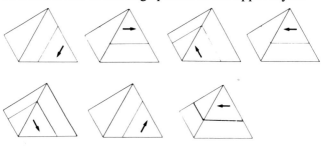

*Note:* Sometimes it happens that *two* remaining edge pieces are correctly positioned when the upper layer is turned. If this occurs, perform the turning sequence described above. This will displace the two edge pieces while correctly positioning the third edge piece.

Then repeat the turning move sequence again. This should result in all three of the edge pieces being correctly positioned and oriented. If it does not, repeat the sequence a third time. There may be other occasions when it is necessary to perform the turning move sequence more than once in order to correctly position all of the edge pieces.

## Special Case

If after completing steps 1-3 your pyramid looks like the one shown below, there is a special turning move sequence you can perform to complete the entire upper layer at once.

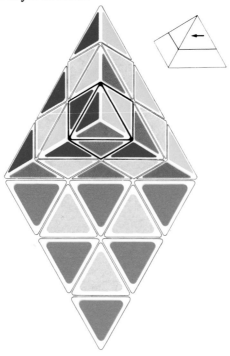

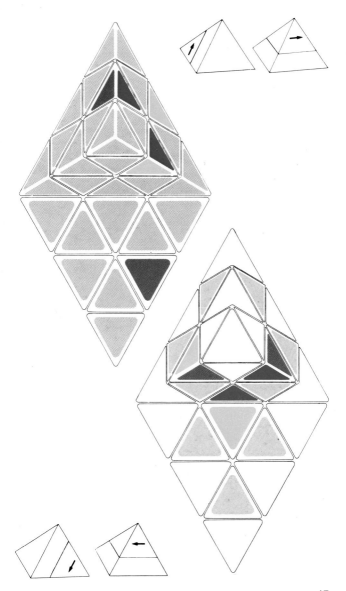

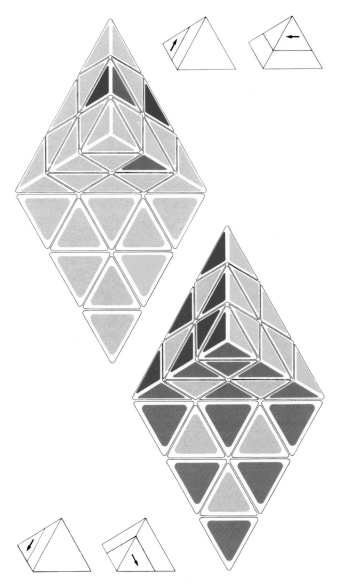

49

Here is the sequence again without the color illustrations:

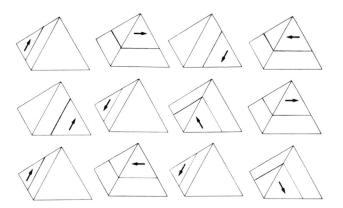

## Stage 3: Positioning the Middle Pieces

The dark grey areas in the illustration below show the three triangular surfaces of what is actually *one solid piece* called a *middle piece*. The three white areas show the three surfaces of another middle piece.

Hold your pyramid so that you are looking straight down at one of the peaks. You will see the three colored surfaces of the middle piece directly below it.

There are four middle pieces in the pyramid, one immediately adjacent to each peak.

Turn your pyramid and carefully examine all four of the middle pieces. Note their relationships to the edge pieces already positioned. At this stage, there are five possibilities for the middle pieces:

**1.** All four are wrongly positioned.

In other words, their colors do *not* match those of the adjacent edge pieces.

**2.** Two are wrongly positioned and two are correctly positioned.

In other words, the colors of two middle pieces match those of the adjacent peaks.

**3.** Three are wrongly positioned and one is correctly positioned.

In other words, the colors of only one middle piece match those of the adjacent peaks.

**4.** One is wrongly positioned and three are correctly positioned.

**5.** All four are correctly positioned. (If this is the case, your pyramid is solved – which should be obvious!)

Determine which of these five possibilities is true for your pyramid. Then proceed to the appropriate solution sequence below.

## 1. All four wrong

Hold your pyramid so that you can easily manipulate the upper layer and the right front layer. You will be correctly positioning both the top middle piece *and* the right middle piece at the same time.

Look carefully at the top middle piece. Now imagine that you are able to move the entire middle piece clockwise or counterclockwise without disturbing any of the adjacent pieces.

If one such move *clockwise* would correctly position the top middle piece – as in the illustration below –

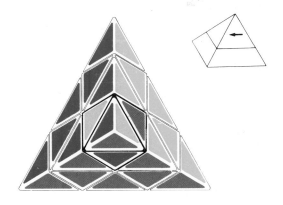

then you will turn the upper layer *counterclockwise* in the turning sequence.

If one such move *counterclockwise* would correctly position the top middle piece – as shown below –

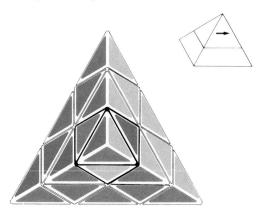

then you will turn the upper layer *clockwise* in the turning sequence:

Once you have determined the direction in which you will be turning the upper layer, do the same for the right front layer. Remember that you will be moving the right front layer in the *opposite* direction from what you would expect, just as you will be doing with the upper layer.

**a** If you will be turning the upper layer clockwise and the right front layer clockwise, perform the following turning move sequence five times:

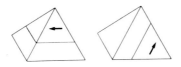

**b** If you will be turning the upper layer counter-clockwise and the right front layer counter-clockwise, perform the following turning move sequence five times:

**c** If you will be turning the upper layer clockwise and the right front layer counterclockwise perform the following turning move sequence five times:

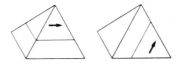

**d** If you will be turning the upper layer counter-clockwise and the right front layer clockwise, perform the following turning move sequence five times:

Your pyramid is now half solved! To complete the solution, turn your pyramid so that the remaining two incorrectly positioned middle pieces are in the upper layer and the right front layer.

Determine the turning direction for each layer, as described above, and perform the appropriate turning sequence.

Reposition any peaks that have been moved out of place.

Your pyramid is now solved!

## 2. Two wrong, two right
Turn your pyramid until the incorrectly positioned middle pieces are in the upper layer and the right front layer.

Solve your pyramid according to the instructions given above.

## 3. Three wrong, one right
Turn your pyramid until two of the three incorrectly positioned middle pieces appear in the upper layer and the right front layer.

Follow the instructions given in **1** above to correctly place the two middle pieces.

Turn your pyramid so that the last remaining incorrectly placed middle piece is in the upper layer.

Imagine again that you are able to move the entire

middle piece clockwise or counterclockwise without disturbing any of the adjacent edge pieces.

If one such move *clockwise* would correctly position the top middle piece, then you will turn the upper layer *clockwise* in the turning sequence.

If one such move *counterclockwise* would correctly position the top middle piece, then you will turn the upper layer *counterclockwise* in the turning sequence.

**a** If you will be turning the upper layer clockwise, your turning sequence will consist of the following *two* sets of moves:

First, repeat this sequence five times:

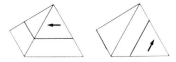

Next, repeat this sequence five times:

Once you reposition the peak, your pyramid will be solved.

**b** If you will be turning the upper layer counter-clockwise, your turning sequence will consist of the following *two* sets of moves:

First, repeat this sequence five times:

Next, repeat this sequence five times:

Reposition the peak to complete the pyramid.

## 4. One wrong, three right
Solve your pyramid according to the instructions given for correcting the single wrongly positioned middle piece.

# Chapter 3:
# Summary of Solution

Following is a summary of the strategy used when solving the Pyraminx.

**1.** Choose your color – red, green, yellow, or blue.

**2.** Correctly position the peaks to determine the four faces of the pyramid.

**3.** Correctly position the edge pieces on the bottom face (the face that contains your chosen color).

**4.** Correctly position any remaining edge pieces.

**5.** Correctly position the four solid middle pieces.

Be sure to reposition the peaks throughout the solution sequence.

# Chapter 4:
# Alternative Sequences

As you become more expert at manipulating and solving the pyramid, you may wish to improve your speed by means of the following alternative sequences.

# 1. Turning the edge pieces on the bottom face one position in a clockwise direction

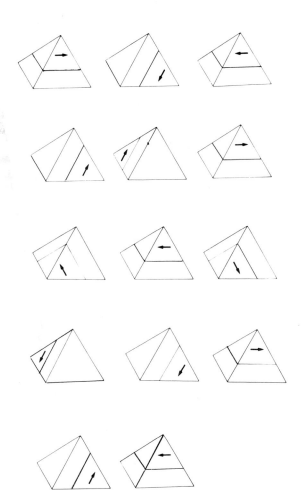

**2. Turning the edge pieces on the bottom face one position in a counterclockwise direction**

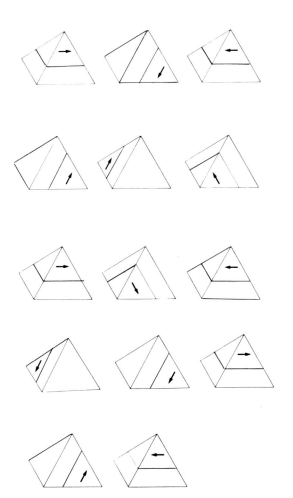

## 3. Turning a middle piece one position in a clockwise direction

In Chapter 1, a logical and easy-to-remember turning move sequence was given for repositioning a single middle piece.

The following turning move sequence is shorter but not as simple to memorize. Once you have mastered it, however, you should be able to cut your turning time by a third or even more.

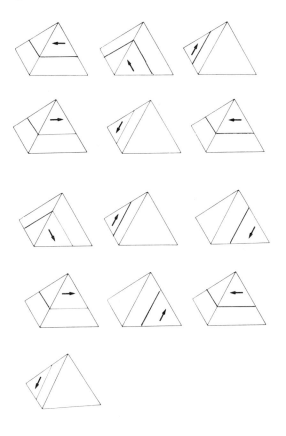

## 4. Turning a middle piece one position in a counterclockwise direction

This is the mirror image of the preceding abbreviated turning sequence.

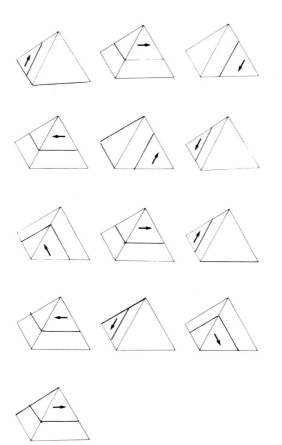

# Chapter 5: Pyraminx Patterns

Once you have mastered the pyramid, you will no longer be satisfied with creating order out of multi-colored chaos. As soon as you understand the basic turning move sequences, you can start experimenting with symmetrical and asymmetrical patterns. Twelve are presented in this chapter.

*Note:* Before beginning each pattern, your pyramid must have solid colors on all four faces.

For each pattern, the red face is the front face and the yellow face is the bottom face.

*Author's Note:* I wish to extend my sincere thanks to Dr. Ronald F. Turner-Smith of the Chinese University of Hong Kong for his helpful advice on the preparation of this book, most notably this chapter.

# 1. Fabian's Windmill

This was the first pattern produced by my youngest son, after whom it was named.

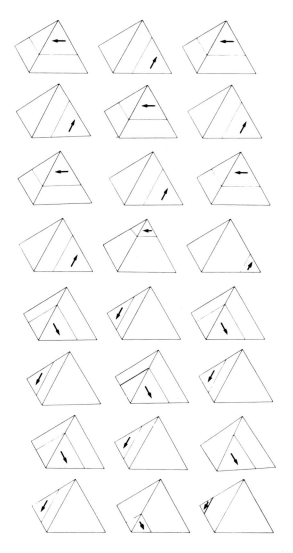

75

## 2. Till's Windmill
This sequence was named after my eldest son. He changed four moves in his younger brother's pattern and deleted four to come up with something entirely different!

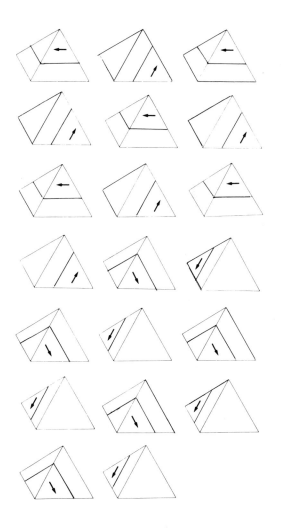

## 3. The Jewel

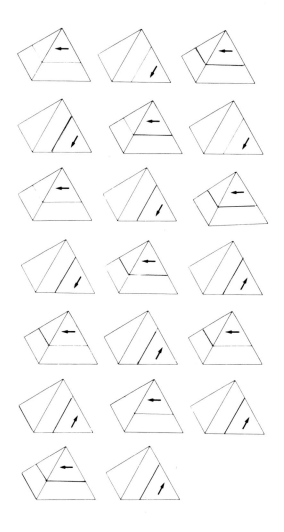

# 4. The Holy Temple Gates

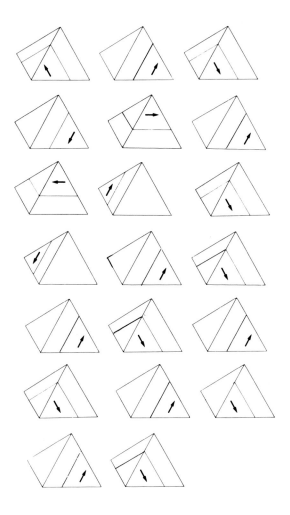

## 5. The Fire-Red Cat's Paw
The cat was a sacred animal in ancient Egypt. The pattern of the cat's paw with outstretched claws is fairly easy to produce.

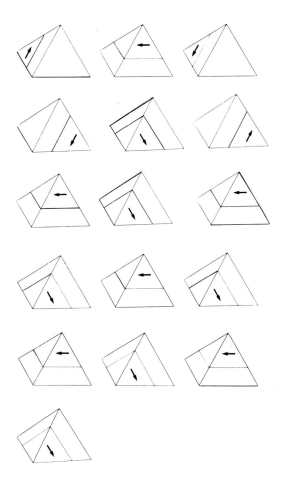

## 6. Blue Heaven over Red Stone

If you turn your pyramid forward slightly after completing this pattern, you will see the blue sky arching over the yellow desert.

The turning sequence for this pattern is exactly the same as the one given for the Jewel, except that in the last move only the peak is turned.

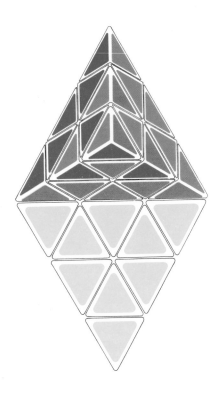

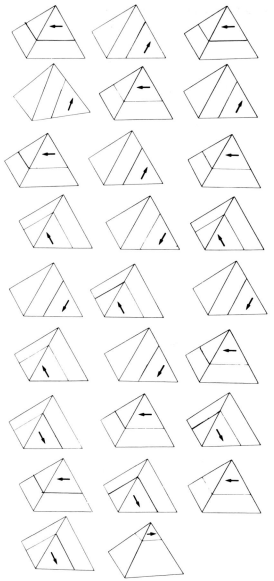

## 7 Tutankhamen's Treasure Chest
This pattern is obtained by repeating the turning move sequence for the Fire-Red Cat's Paw a total of three times.

For the first turning move sequence, the red face is the front face and the yellow face is the bottom face.

For the second sequence, the green face is the front face and the red face is the bottom face.

For the third sequence, the yellow face is the front face and the green face is the bottom face.

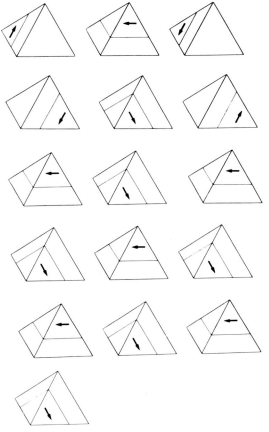

## 8. Nefertiti's Jewel Box
This colorful pattern is a variation on Tutankhamen's Treasure Chest.

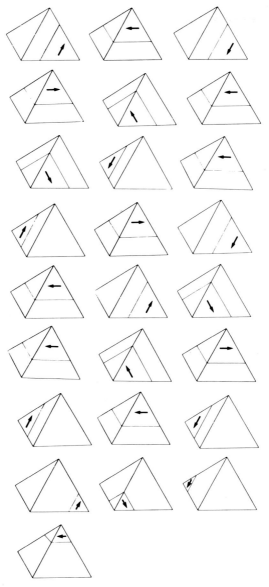

## 9. Cleopatra's Needle

This pattern is a variation on one of the preceding patterns. Try to guess which!

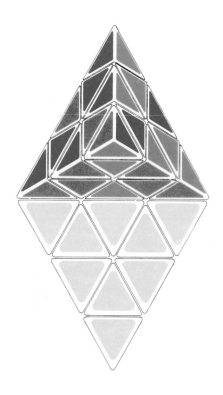

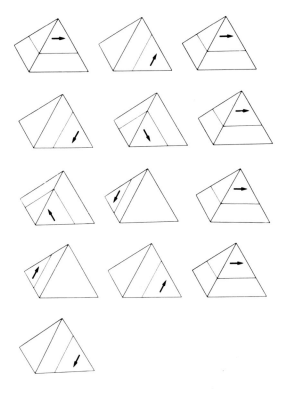

# 10. The Wagon Wheel

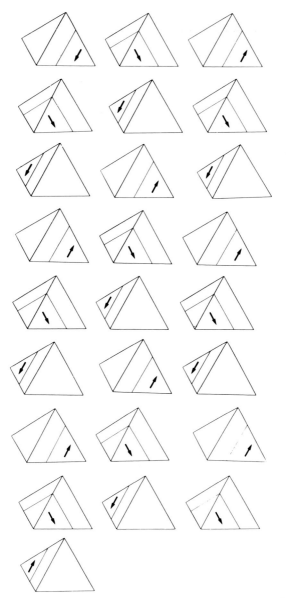

# 11. The Whirlwind

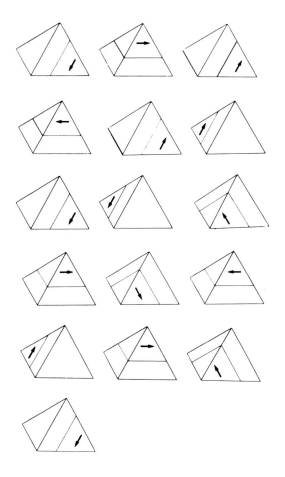

## 12. The Philosopher's Stone
Three pillars of fire guard the gold, sapphire, and emerald scepters lying crossed beneath the pyramid.

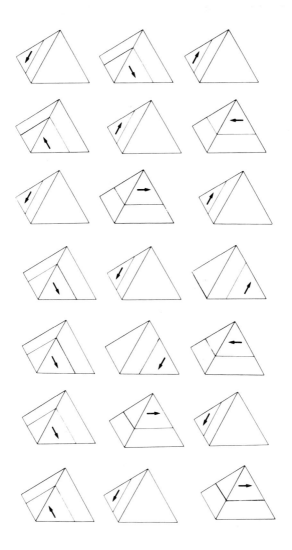

## Three Impossible Patterns

Every pyramid aficionado sooner or later tries to create one or more of these patterns. Although they all look quite logical, no amount of turning can achieve them!

## The Riddle of the Sphinx

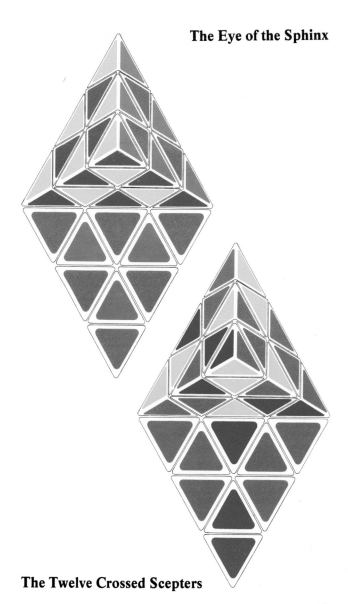

**The Twelve Crossed Scepters**

# Chapter 6:
# The Pyraminx for the
# Visually Impaired

Uwe Mèffert, the inventor of the Pyraminx, has also designed a version of the pyramid puzzle that may be used by the visually impaired. He experimented with a number of self-adhesive tactile materials and tested them widely before arriving at the virtually trouble-free pyramid shown below.

For information, price, and availability, contact the following:

David Singmaster Ltd
66 Mount View Road
London NY 4JR
England

Telephone: 01-348 7266

Tomy Corp.
901 E. 233 St.
Carson, California 90745
U.S.A.

# Afterword:
# The Story of the Pyraminx

The ancient Greek philosophers perceived earth, fire, air, and water as elements consisting of atoms. According to them, these atoms had distinct geometric shapes. Earth atoms were cubes; fire atoms were equilateral pyramids; air atoms were octahedrons; water atoms were icosahedrons (20-sided polyhedrons). Finally, dodecahedrons (12-sided polyhedrons) surrounded the world.

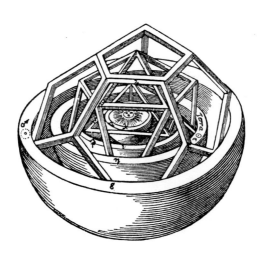

People have always been fascinated by basic geometric forms. One reason why Rubik's Cube has been so successful is that its shape is both obvious and familiar. Similarly, the triangular surfaces of the Pyraminx are utterly symmetrical and therefore soothing. Solving the pyramid puzzle, however – arranging its surfaces in such a way that each contains a solid color – can be rather exasperating. Most people who play with the puzzle are unable to rest until they have either penetrated its mysteries or at least come one step closer to the solution.

Clearly, there are similarities between the cube and the pyramid. Both are three-dimensional and have brightly-colored surfaces. Both are constructed in startingly simple yet ingenious ways. Both are intellectually challenging and stimulating. Yet it would be wrong to link them too closely in your mind. In fact, it is best to forget about the cube while you are working on the pyramid. The latter requires a completely different perspective and set of techniques.

All of the planes of the cube may be moved either horizontally or vertically; the sections of the pyramid, on the other hand, may be moved either horizontally or at an angle. The cube has six sides, the pyramid four. Thus each puzzle has its own individual character and method of solution.

It should be said that the pyramid was not conceived in the wake of the cube. Uwe Mèffert first constructed a fully-functioning wooden prototype nearly ten years ago, in the early seventies. Once he was satisfied that it worked, he simply put it away and went on to other things.

Uwe Mèffert is of German-French origins. He grew up in Heidelberg, went abroad, married, and had travelled the world before he finally settled in Hong Kong. His intellect is formidable, and he has had an impressive array of experiences. For nearly 20 years he has been involved in the field of psychology. He has worked as a consultant in a number of areas, but prefers to consider himself an inventor rather than an advisor. He has, in fact, developed and produced medical equipment aimed at improving physical fitness. He is also an advocate of acupuncture and its healing properties.

During his travels, Mèffert visited the enormous pyramids of Giza and became intensely interested in them. He saw a relationship between acupuncture and the quiet power symbolized by the pyramid shape. The original pyramid – the one Mèffert made almost a decade ago – was first intended as a therapy tool. Rubbing the lightly-rounded peaks of a small pyramid into the palm of the hand has a calming effect similar to that produced by an acupuncture relaxation exercise.

I have known Mèffert for several years, and we have a number of interests in common. Interestingly enough, we are even the same age. We both like to occupy ourselves with mathematical/logical problems and puzzles. When Mèffert realized that a book about solving his pyramid was needed, he turned to me, and I was more than happy to write it.

The pyramid may have lain dormant in Mèffert's drawer forever had it not been for the enormous popularity of the cube. This wonderful little toy has broken all records for any game ever produced. Monopoly™ has sold 80 million in about 45 years; Scrabble™, 45 million in 34 years; Mastermind™, 30 million in 10 years. In contrast, the cube has sold 50 million in about one year.

Fortunately, Mèffert did not rush blindly into production when he decided to market his pyramid. He knew that mistakes had been made with the cube, and he did not wish to duplicate them. The inventor of the original cube, the Hungarian professor Ernö Rubik, had not taken sufficient pains to protect his patent rights. Cheap imitations – many of very poor quality – sprang up almost overnight.

Mèffert wanted to guarantee that his pyramids would be of consistently high quality. The only way to do this was by securing worldwide patents. This he accomplished, and named his construction the Pyraminx.

Since that time Mèffert has influenced all aspects of the production process. The pyramids are made from ABS, an extremely tough and break-resistant plastic. (Incidentally, the same plastic will one day be used in the manufacture of telephones.) The ball-bearing mechanism allows the surface to be turned smoothly and rapidly, and the stops hold the planes firmly in place in between moves.

Although the pyramid is brilliantly simple, it is quite difficult to produce. This is because much of the human environment is based on the right angle. Angles of 60° and rotations of 120° – both of which are integral to the pyramid – are unfamiliar and strange. This at first caused problems. After some experimentation, Mèffert arrived at an intricate solution.

At the center of the pyramid is a ball with four holes. These holes point to the four peaks of the pyramid. The ball is connected to the four middle pieces of the pyramid by means of threaded screws inserted into plastic tubes. A small movable triangle is placed at the head of each screw. In the final steps of

the assembly, a triangular cap is attached to each movable triangle to form the peak.

The six edge pieces are fitted into small grooves in the middle pieces. Finally, color patches are stuck to the thirty-six separate triangular surfaces of the pyramid.

This has proved to be the most functionally efficient means of construction. Nonetheless, Mèffert has protected his patent with respect to other forms of construction that have essentially similar mechanisms and movements – including one in which the pieces are held together magnetically.

Mèffert has continued to develop other mind-challenging games. He has now made several pyramids ranging in complexity. For one variation, Mèffert changed each triangular surface into a little equilateral pyramid. The resulting prickly structure, while appearing more complex than the smooth one, is actually easier to solve.

Two more versions will be available in the future. These will be similar in outward appearance to the original Pyraminx but will have different internal mechanisms. Solving them will thus require new (and significantly more difficult) movement sequences.

In practice it has been found that the solution for the various pyramids require the following *minimum* number of steps:

| | |
|---|---|
| The "prickly" pyramid | 24-28 steps |
| The standard Pyraminx, the solution for which is presented in this book | 38 steps |
| Pyramids which are not yet on the market but are covered by patents and are ready for production | 215-255 steps |

It is possible, of course, that the minimum number of moves needed to solve the pyramid may be found to be fewer than 38. However, it is doubtful that this number will be reduced by much.

The standard Pyraminx is not quite as difficult to master as the cube, which requires a minimum

number of 50 moves to solve. Most people, in fact, find it impossible to work the cube through to completion without detailed instructions. While the pyramid is just difficult enough to torment you for a long period of time, it is possible that you can solve it on your own.

If after a considerable struggle you do manage to figure out the pyramid by yourself, you don't need to stop there. The pyramid is full of mysteries and possibilities. As you grow in your understanding of it, you will discover even more ways to approach

*Opposite: The prickly pyramid*
*Above: A clear plastic pyramid revealing the inner mechanism*

and enjoy it. And you may never see the same starting pattern in a lifetime.

It has been shown mathematically that there are over 75 million possible starting patterns – 75,582,720, to be exact. For Mèffert's Master Pyraminx (which is not yet available) there are 446,965,972,992,000 possible starting patterns! The Master Pyraminx will prove to be quite a challenge, even for puzzle experts.

This is the first book to be written about the

Pyraminx. As yet there is no one who has emerged as a recognized specialist on this amazing game. There are no clubs which devotees may join, no publications to which they may subscribe, no anecdotes to record and share. In many ways, then, this book is only the beginning of the pyramid's story.

**'It was uncomplicated and great— at least that's what you told me at the time. But now you're using that fact to make me sound like some sort of *tramp*!'**

Steeling himself against the venom spitting from Catrin's eyes, Murat shook his head. 'My intention has never been to make you feel bad about yourself, but I never promised you marriage, Cat. I made that clear from the start. And, yes, I always said I would be honest with you. I told you that I had very particular specifications for my bride. That I required a woman of royal blood who would behave as a future desert queen would be expected to behave. And I'm afraid that you were…'

His voice tailed away, but once again she didn't seem prepared to let it go.

'I was *what*, Murat?'

He sighed, wondering why women always did this. Even Cat. Why they always provoked you to a certain point and made you say something which afterwards you would both regret. He shook his head. 'It doesn't matter.'

'Oh, but it does. It matters very much.'

He shrugged.

# DESERT MEN OF QURHAH

*Their destiny is the desert!*

The heat of the desert is nothing compared to the passion
that burns between the pages of this stunning new trilogy
by Sharon Kendrick!

## Defiant in the Desert
December 2013

Oil baron Suleiman Abd Al-Aziz has been sent to
retrieve the Sultan of Qurhah's reluctant fiancée—
a woman who's utterly forbidden, but is determined to
escape the confines of her engagement…by seducing him!

## Shamed in the Sands
February 2014

The Princess of Qurhah has always wanted
something different from her life. So when sexy
advertising magnate Gabe Steele arrives to work for
her brother, Leila convinces Gabe to give her a job…
but that's not the only thing to cause a royal scandal!

## Seduced by the Sultan
April 2014

The Sultan of Qurhah is facing a scandal of
epic proportions. His fiancée has run off,
leaving him with a space in his king-sized bed.
A space once occupied by his mistress—Catrin Thomas.
And now he wants her back—at any price!